BRITANNICA
Mathematics in Context

Made to Measure

Britannica
ENCYCLOPÆDIA BRITANNICA EDUCATIONAL CORPORATION

Mathematics in Context is a comprehensive curriculum for the middle grades. It was developed in collaboration with the Wisconsin Center for Education Research, School of Education, University of Wisconsin–Madison and the Freudenthal Institute at the University of Utrecht, The Netherlands, with the support of National Science Foundation Grant No. 9054928.

National Science Foundation

Opinions expressed are those of the authors
and not necessarily those of the Foundation

ISBN 0-7826-1525-2
1 2 3 4 5 WK 02 01 00 99 98

The *Mathematics in Context* Development Team

Mathematics in Context is a comprehensive curriculum for the middle grades. The National Science Foundation funded the National Center for Research in Mathematical Sciences Education at the University of Wisconsin–Madison to develop and field-test the materials from 1991 through 1996. The Freudenthal Institute at the University of Utrecht in The Netherlands, as a subcontractor, collaborated with the University of Wisconsin–Madison on the development of the curriculum.

The initial version of *Made to Measure* was developed by Jan de Lange and Monica Wijers. It was adapted for use in American schools by Aaron N. Simon and Mary C. Shafer.

National Center for Research in Mathematical Sciences Education Staff

Thomas A. Romberg
Director

Joan Daniels Pedro
Assistant to the Director

Gail Burrill
Coordinator
Field Test Materials

Margaret R. Meyer
Coordinator
Pilot Test Materials

Mary Ann Fix
Editorial Coordinator

Sherian Foster
Editorial Coordinator

James A. Middleton
Pilot Test Coordinator

Project Staff

Jonathan Brendefur
Laura J. Brinker
James Browne
Jack Burrill
Rose Byrd
Peter Christiansen
Barbara Clarke
Doug Clarke
Beth R. Cole
Fae Dremock
Jasmina Milinkovic
Margaret A. Pligge
Mary C. Shafer
Julia A. Shew
Aaron N. Simon
Marvin Smith
Stephanie Z. Smith
Mary S. Spence

Freudenthal Institute Staff

Jan de Lange
Director

Els Feijs
Coordinator

Martin van Reeuwijk
Coordinator

Project Staff

Mieke Abels
Nina Boswinkel
Frans van Galen
Koeno Gravemeijer
Marja van den Heuvel-Panhuizen
Jan Auke de Jong
Vincent Jonker
Ronald Keijzer
Martin Kindt
Jansie Niehaus
Nanda Querelle
Anton Roodhardt
Leen Streefland
Adri Treffers
Monica Wijers
Astrid de Wild

Table of Contents

Dear Student,

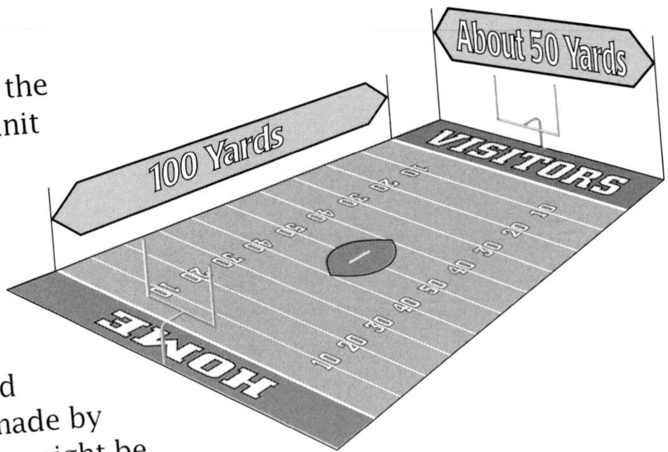

You will soon be starting the *Mathematics in Context* unit *Made to Measure*. The unit is all about measuring: measuring your feet, hands, and thumbs; measuring wood, a football field, and a deck of cards; and measuring the angles made by your arm and wrist. You might be amazed by what you can measure!

You will find that mathematics plays an important role in measurement. Every time you measure something, you might ask yourself the following questions:

- Will every person measuring this item get the same measurement that I did?
- Do all of these things have the same measurement?
- What other units of measure can I use?
- Are there other ways to measure these things?

Whenever you make a measurement in this unit, picture how big—or small or steep or short—that measurement is. When you can do this with all of the measurements in this unit, you are well on your way to becoming a mathematician!

Sincerely,

The Mathematics in Context Development Team

A. LENGTHS

INTRODUCTION

People who design the objects you use every day have thought a lot about how big or how small those objects should be. Knowing the sizes of people's arms, legs, and hands can be very useful when designing furniture, clothes, toys, windows, doors, and many other items.

1. Which body measures would be useful to know if you were designing the following:

a. doors

b. school desks

c. shoes

d. pants

e. baby cribs

f. stairs

2. For what other objects would you need to know body measures?

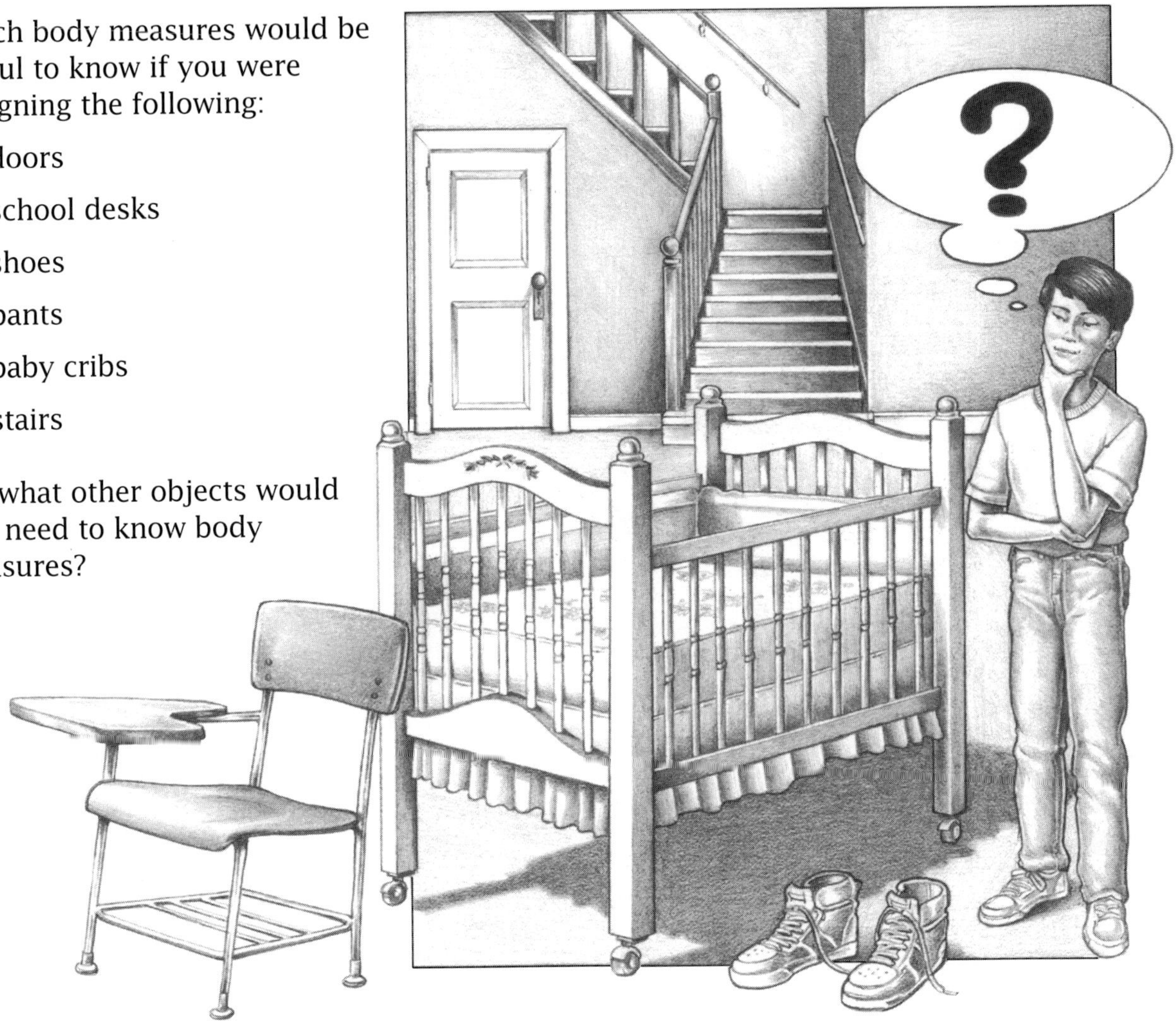

HISTORICAL MEASURES

You probably discovered that the lengths and heights of different body parts are important for designing many common objects. At one time, all measurements for length were related to the human body. Some of these units of measure include the *thumb, hand span, foot, yard, pace,* and *fathom.*

3. Match each of the units of measure listed in the above paragraph with its drawing on page 2.

a.

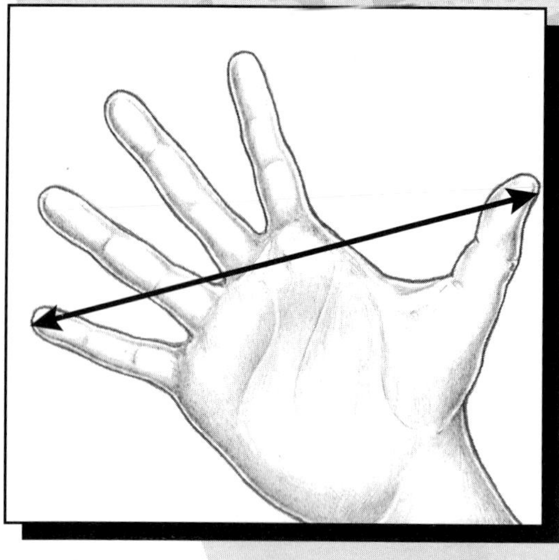

b.

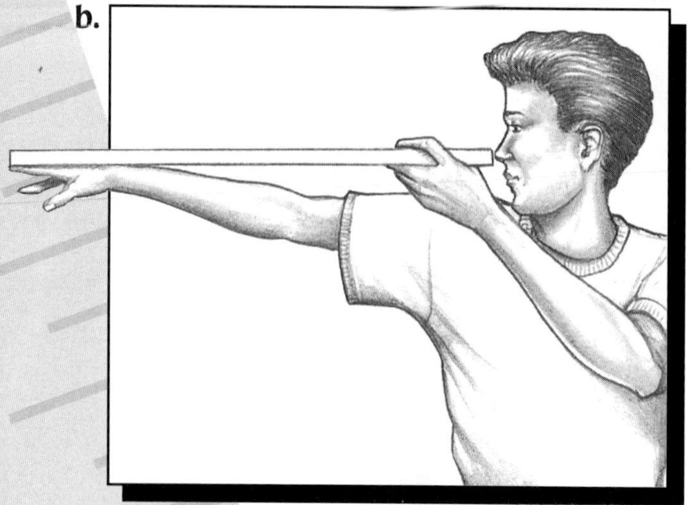

c.

d.

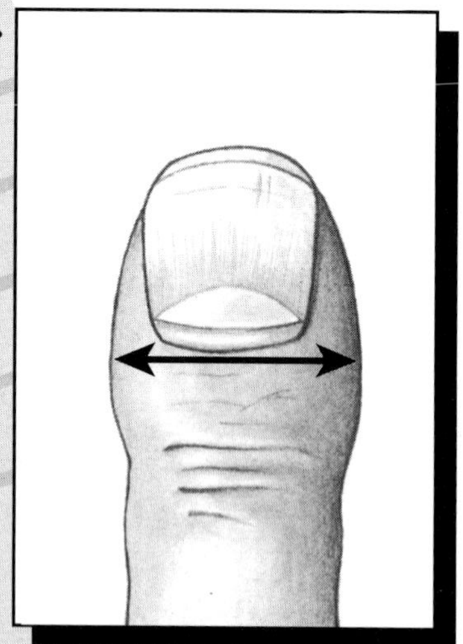

e.

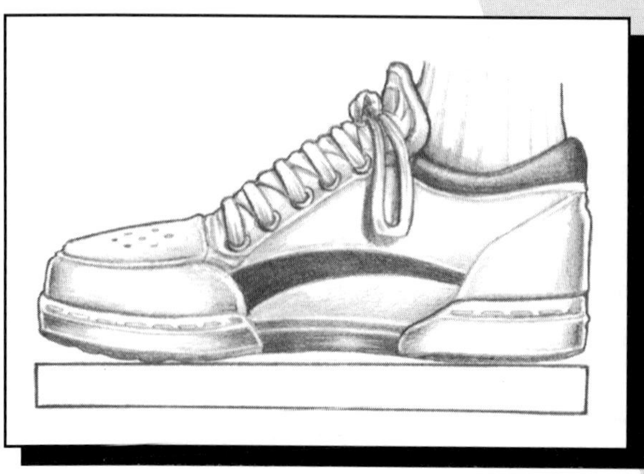

f.

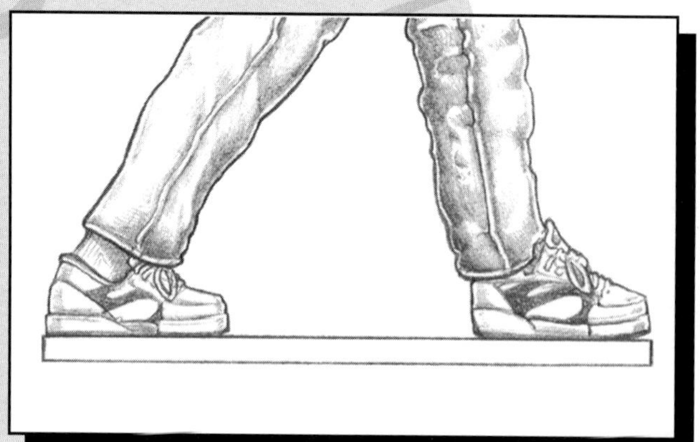

4. a. Which of the units of measure from problem **3** would you use to find the length of the nail shown on the right?

b. How long is this nail in the units of measure you chose for your answer to **a?**

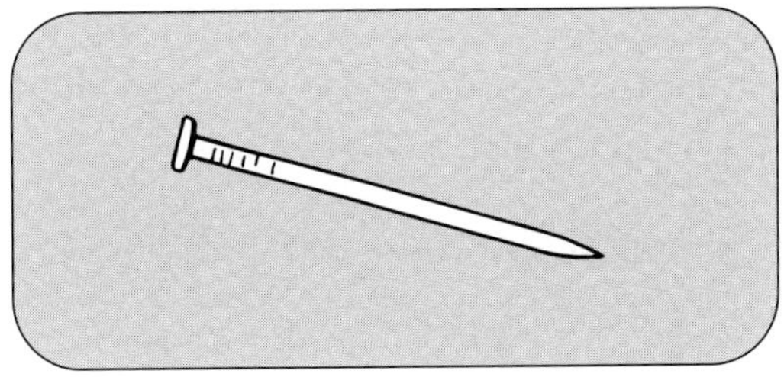

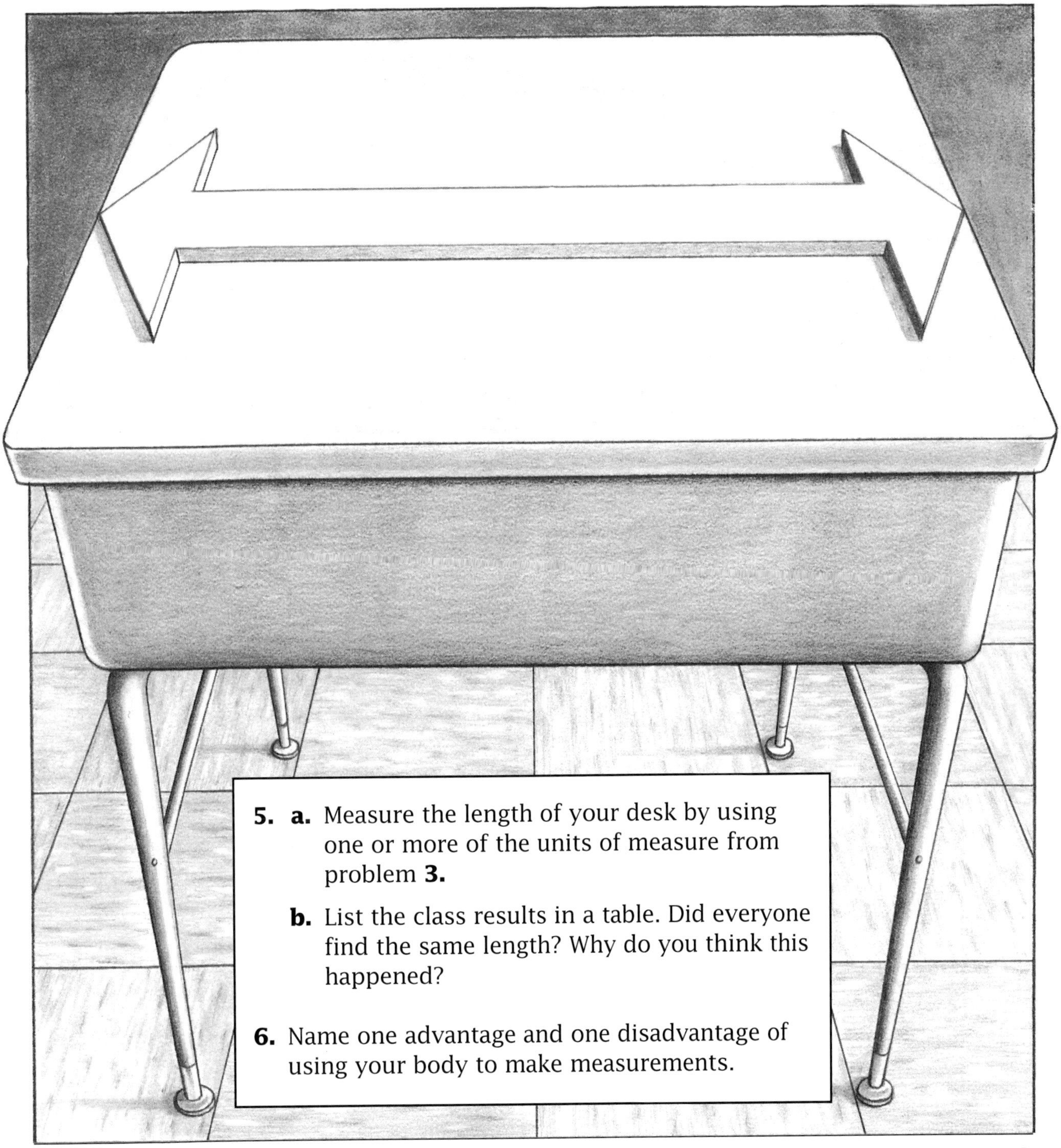

5. a. Measure the length of your desk by using one or more of the units of measure from problem **3.**

b. List the class results in a table. Did everyone find the same length? Why do you think this happened?

6. Name one advantage and one disadvantage of using your body to make measurements.

In Scotland during the Middle Ages, a unit of measure called the Scottish thumb was used. A *Scottish thumb* is the mean of the thumb widths of three men: a large man, an ordinary sized man, and a small man.

7. Why were three different-sized men used to determine the Scottish thumb?

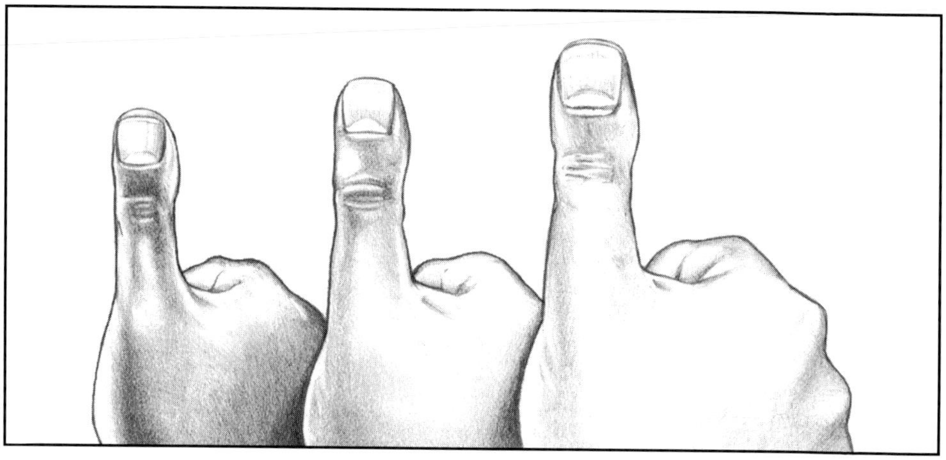

In 1616, the Germans decided to create a unit of measure called the *mean foot.* To do this, they cut a piece of rope that was as long as the feet of 16 men.

8. **a.** How do you think the rope was used to find the length of the mean foot?

b. Which measurement is closer to the average person's measurement: the German mean foot or the Scottish thumb? Explain your answer.

9. With the help of 16 classmates, find the length of the mean foot in your class by using the method described above.

10. Measure the following in centimeters. List the results in a table.

a. your thumb width

b. your hand span

c. your foot length

d. your pace

11. Use the table you made in problem **10** to find the following relationships between measures:

a. How many thumb widths are in one hand span?

b. How many thumb widths are in one foot?

c. How many feet are in one pace?

You may add other units of measure and the relationships between them to your list.

12. What is the size of the "typical" thumb for your classmates? Explain your answer.

Imagine what it must have been like when everyone used their own thumbs for measuring. Today, of course, we have standard systems of measurement.

A few countries, including the United States, still use the foot as a unit of measure, but the length of a foot no longer refers to the length of each person's foot. A standard has been officially established for the length of a foot. Most countries use the *metric system*, which was adopted in France in 1795.

13. a. The foot is a part of the English system of measurement. List some other units of measure for length that are part of the English system.

b. In your notebook, write as many relationships between the units of measure in the English system as you can.

14. a. List some units of measure for length that are part of the metric system.

b. Write as many relationships between these units of measure as you can.

15. Find some simple relationships between the English and metric systems.

Feet and Shoes

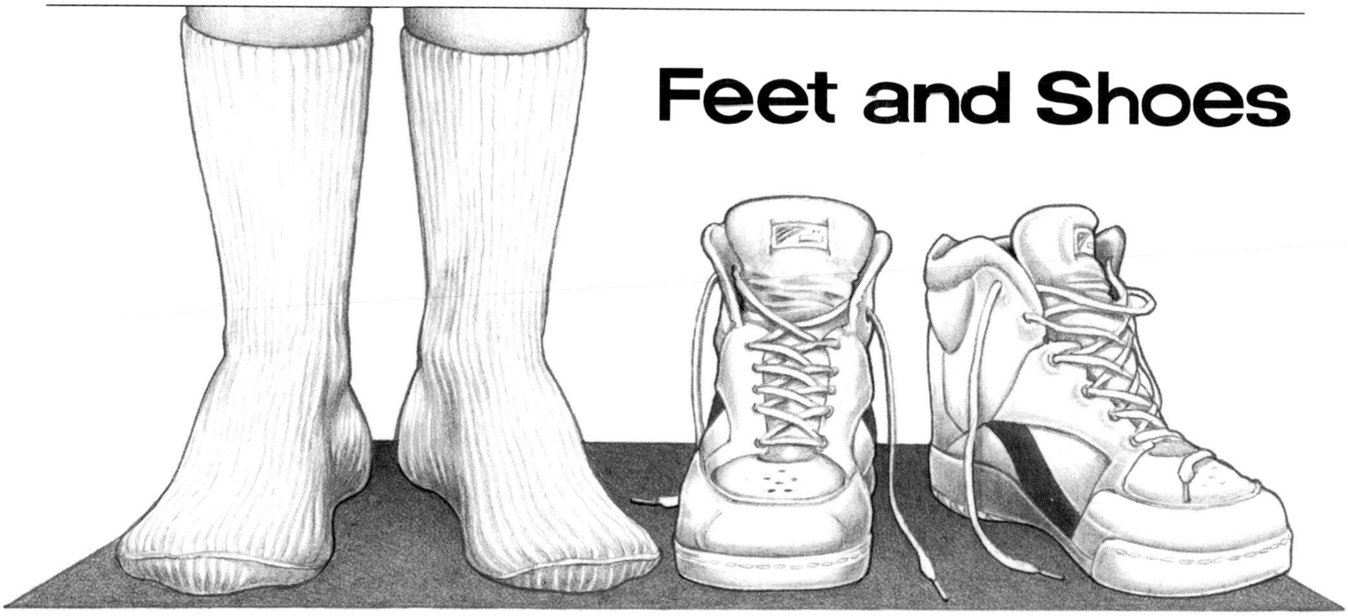

In problem **10c,** you measured the length of your foot in centimeters. The length of your foot is different from your shoe size.

16. a. Do you think that there is a relationship between shoe size and foot length? Explain your answer.

b. Make a table that lists the foot length (in centimeters) and corresponding shoe size for each student in your class.

c. Graph the results. Put foot length in centimeters on the horizontal axis and shoe size on the vertical axis. What does your graph tell you about the relationship between foot length and shoe size?

Just as countries use different systems of measurement, they also have different systems for determining shoe sizes. For some shoes, you can find at least three different sizes:

- European size—usually a number between 33 and 47
- U.K. (United Kingdom), or English, size—usually a number between 1 and 15
- U.S. size—usually a number between 1 and 15 (slightly larger than U.K. sizes)

The U.K. system of shoe sizes began in the seventh century. Shoe sizes were measured with a standard thumb (now called an inch).

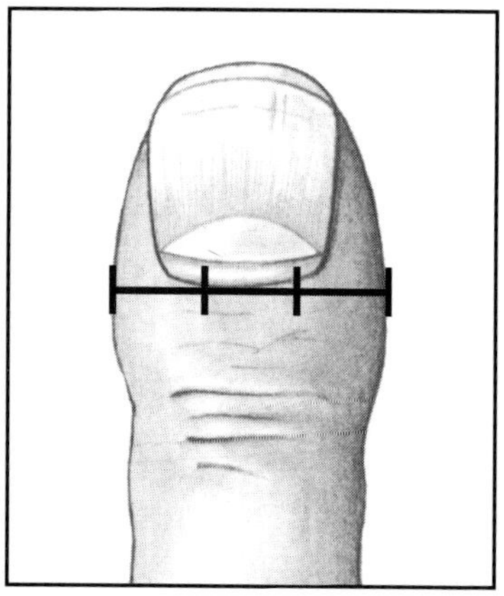

17. How many standard thumbs (or inches) long is your foot?

To get a more accurate measurement, the U.K. introduced a smaller unit of measure, the stitch. Three *stitches* are in one standard thumb.

18. How many stitches long is your foot?

In the U.K. system, shoe sizes are based on the number of stitches. The first 25 stitches are not counted in adult shoe sizes. Size 1 is, therefore, really 26 stitches, or $8\frac{2}{3}$ inches.

19. a. Copy the table below into your notebook and continue it to shoe size 8.

Foot Length (in stitches)	Foot Length (in inches)	United Kingdom Shoe Size
26	$8\frac{2}{3}$	1
27		

b. Use your answer to problem **18** to find your U.K. shoe size in the table you made in part **a.** How does your U.K. shoe size compare with your U.S. shoe size?

c. Formulate a rule that helps you find someone's foot length in inches if you know his or her U.K. shoe size. Write the rule in arrow language.

U.K. shoe size $\xrightarrow{?}$ ____ $\xrightarrow{?}$ foot length in inches

20. After you finish problem **19,** look back at your answers to problem **16.** Would you change your answers now? Why or why not?

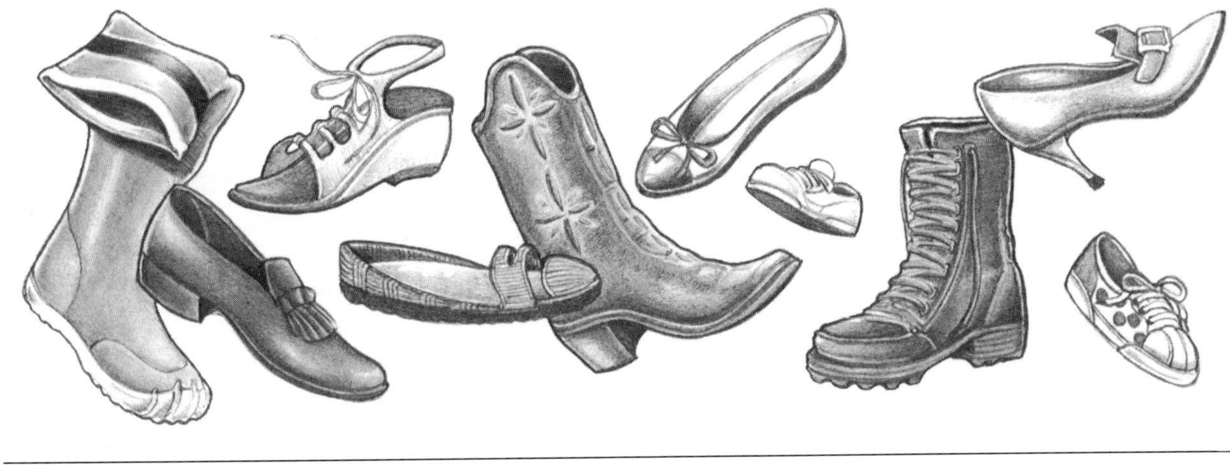

Feet and Miles

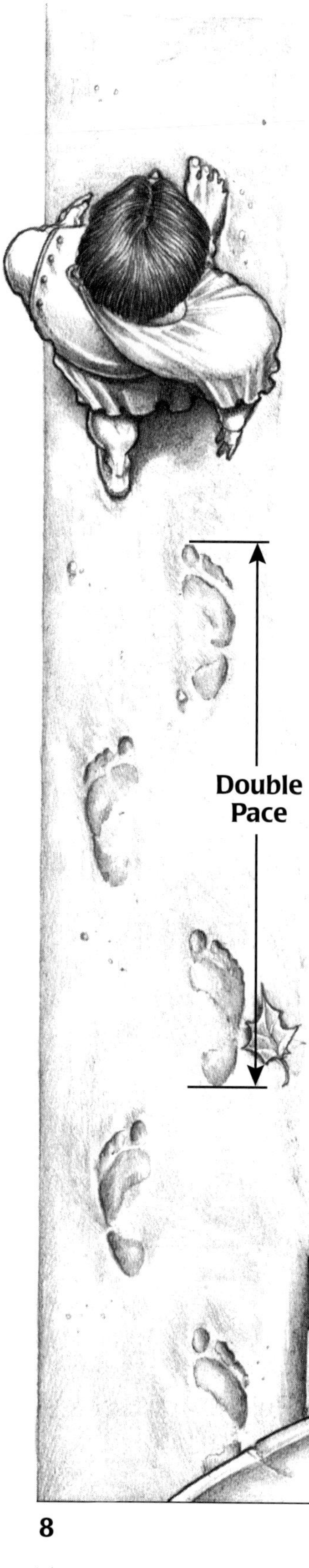

In problem **10** you measured your pace and your foot length.

21. Make a table of the pace and foot-length measurements of all of the students in your class. Use the results to determine whether the two measures are related. For example, you can find out if the person with the biggest foot also has the longest pace or if the person with the shortest foot has the shortest pace. Drawing a graph may be helpful.

The Romans invented the mile as a unit to measure distances. The word *mile* comes from *millia passuum,* which means "1,000 double paces." *Note:* The *double pace* is not two paces long. It is measured from the back of one foot to the front of the same foot after two steps have been taken.

22. Using the Roman definition of a mile being equal to 1,000 double paces, find the length of your own "personal" mile.

In 1592, a British law about a different kind of mile read:

"The statute mile is 8 furlongs, of 40 rods each, of 16.5 feet each."

A *furlong* is another old measure. The word comes from the phrase "a furrow long."

23. What is the meaning of the word "furrow"? Find the length of one furrow by using the above law.

24. **a.** Compare your personal mile with the Roman mile (which is 1,478.7 meters long).

b. Compare your personal mile with the statute mile, which is the one normally used today.

c. Write a paragraph about your results.

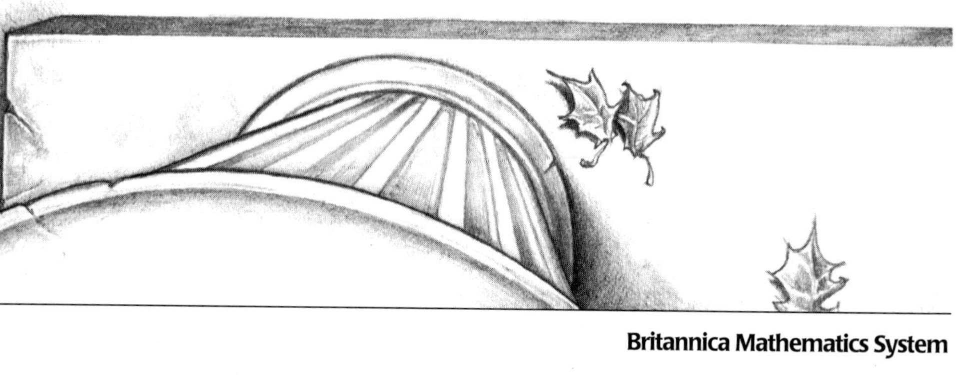

Body Length and Fathom

The fathom is another unit of measure associated with the human body. The *fathom* originated as the distance from the middle fingertip of one hand to the middle fingertip of the other hand of a person holding his or her arms fully extended sidewards.

The picture of a man shown on the right is based on a famous drawing by Leonardo da Vinci. As you can see, the man more or less fits in a square.

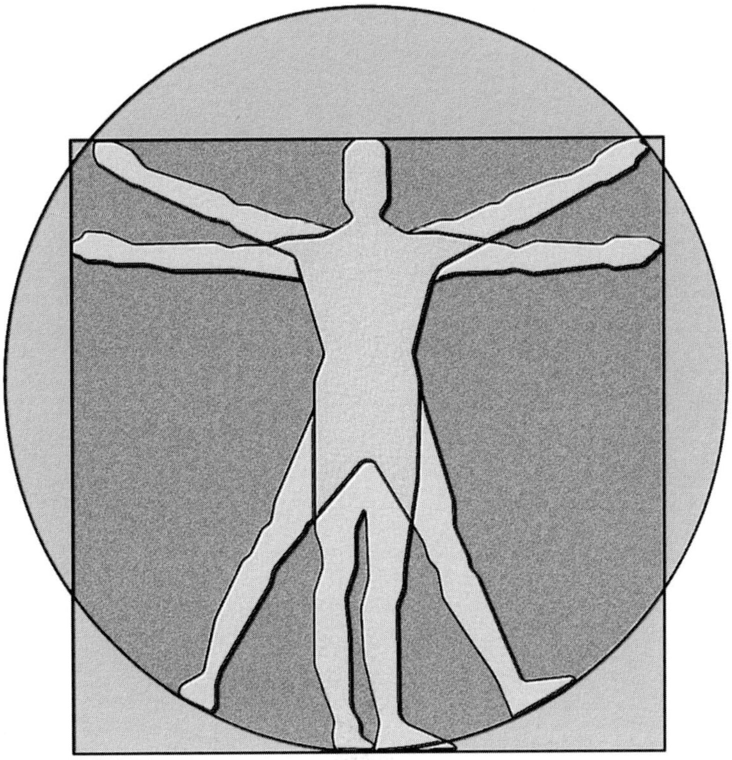

25. Based on the pictures on this page, what is the relationship between a person's height and his or her fathom?

26. Measure your height and your fathom to decide how precisely you would fit into a square.

Other Measures for Length

There are many other ways to measure length. In Papua, New Guinea, for example, a local unit of distance is "a day's travel."

27. Why might a day's travel make sense as a unit of distance?

In mountainous regions in New Guinea, walking distances are expressed in hours, not in kilometers or miles. For example, a citizen of New Guinea might say that the distance from the village to the lake in the mountains is two hours, but the distance back is five quarters of an hour.

28. **a.** Why do you think there are two travel times?

b. Is the distance in kilometers different for the two directions? Explain.

In the 14th century, the biggest trading empire in Africa was the Empire of Mali. Mansa Musa was one of its emperors. Sheik Uthman ed-Dukkali, a learned Egyptian who lived in Mali for 35 years, declared that Mali was "four months of travel long and four months wide."

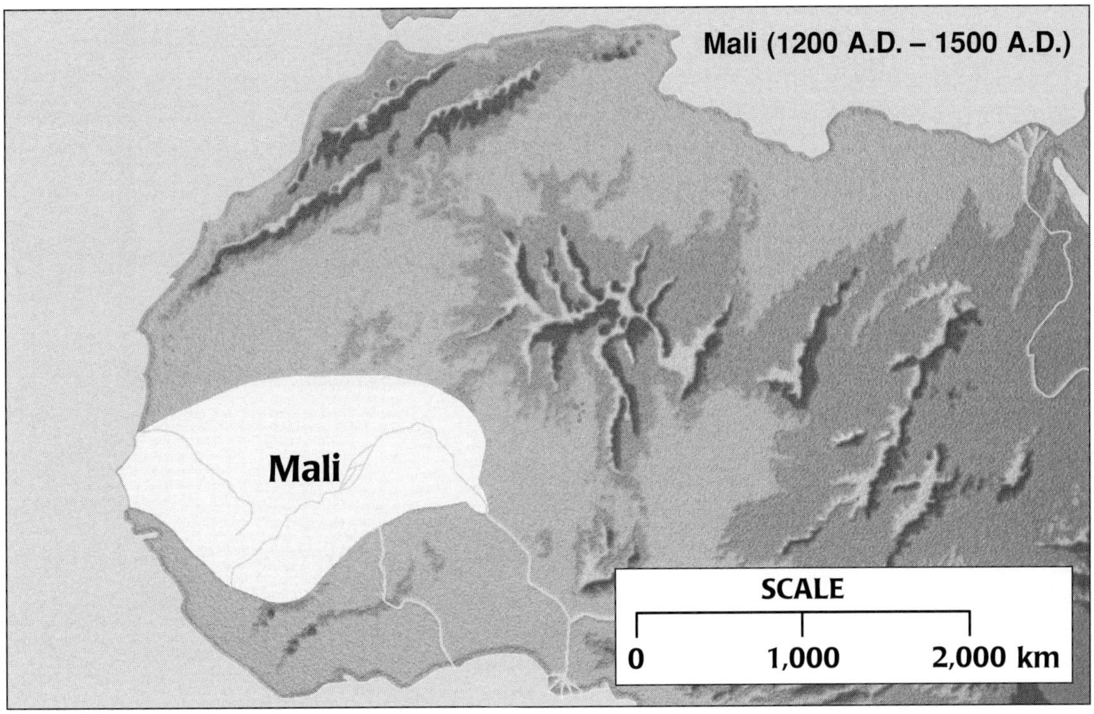

Source: Data from Basil Davidson, *African Kingdoms* (New York: Time Incorporated, 1966).

29. **a.** Use the above picture to estimate the length and width of the Empire of Mali in both English and metric units of measure. (*Note:* 500 miles = 800 kilometers)

b. What is the distance in miles or kilometers of "one month of travel?"

c. Based on your answer to part **b,** how do you think people in Mali traveled in the 14th century?

30. List some distances that are usually expressed in units of time rather than length. Explain why using time is appropriate in each case.

Summary

In this section, you learned many different ways to estimate and measure length. In the past, people used thumbs, feet, and arms to measure length, but the same distance was often measured differently by each person.

Today, two standard systems of measurement exist. Most countries of the world use the metric system, in which length is measured in centimeters, meters, and kilometers. A few countries use the English system, in which length is measured in inches, feet, yards, and miles.

Summary Questions

31. Measure the following lengths any way you like. Discuss your answers in class.

a. the distance from your home to school

b. the length of your classroom

c. the thickness of a penny

B. AREAS

INTRODUCTION

A body's *surface area* is the amount of skin that covers a body. Sometimes it is important to know the surface area of a person's body. For example, health care workers use estimates of a burned patient's body surface area to decide how much liquid is needed to replace the patient's lost fluids.

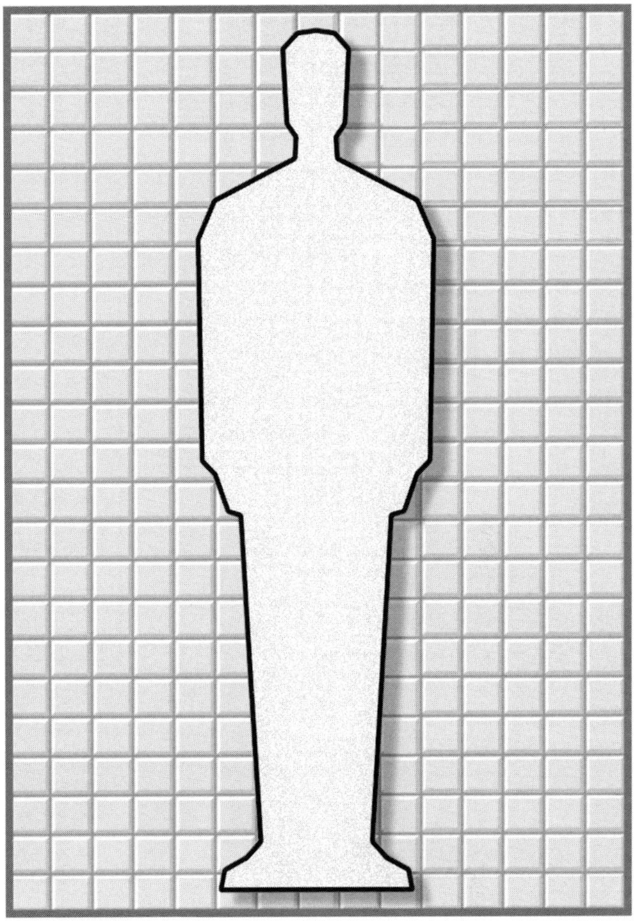

Body surface area is also important in understanding the needs of babies. A body cools down by sweating at a rate proportional to its skin area, but warms up in proportion to its body mass. A baby's skin area is big compared to his or her weight. Since babies cool down faster than adults, babies can feel uncomfortably cold even when adults feel warm.

1. How might you measure your body's surface area (or amount of skin)?

2. Estimate your body's surface area in square centimeters.

As you go through this section, you will estimate your body's surface area in many different ways. Make a table similar to the one shown below to keep track of your estimates.

Surface Area Estimation Table

Estimation Method	Estimation of Surface Area (in cm^2)

Every time you make a new estimate, record your answer in the table.

SQUARES

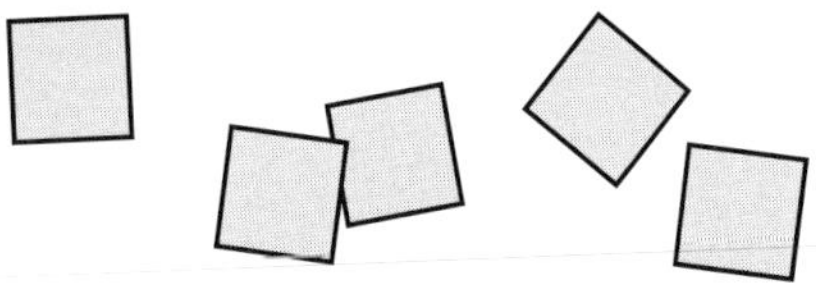

In Section A, you saw how you can fit in a square.

Of course, not everyone's square is the same size. You can use the square in which you fit to approximate your surface area. Three-fifths of your square's area is a good estimate of your surface area.

3. Explain why $\frac{3}{5}$ of the square is a good estimate of your body's surface area.

Activity

4. a. Make a life-size version of your square out of newspaper. (You'll probably need to tape some pages together.) Lie down on one end of the square and roll over one full rotation. Draw a straight line on the paper along the side of your body to mark the end of your roll. Did you roll over about $\frac{3}{5}$ of the square?

b. Shade $\frac{3}{5}$ of your newspaper square. Find the area of the shaded part in square centimeters. Record the result in your Surface Area Estimation Table.

c. To see how close your estimate is, you can try to "put on your skin." Using the newspaper square you made in problem **4a,** try to cover yourself. You may need to cut up your rectangle with scissors. Report how well your "skin" covers you.

There are other ways to find your body's surface area. Timm Ulrichs, a German artist, did many artistic experiments that were quite mathematical. Mr. Ulrichs found his own surface area in the following way:

He took thousands of little sticky squares of foil (exactly the size shown below) and stuck them on his body, until he was completely covered.

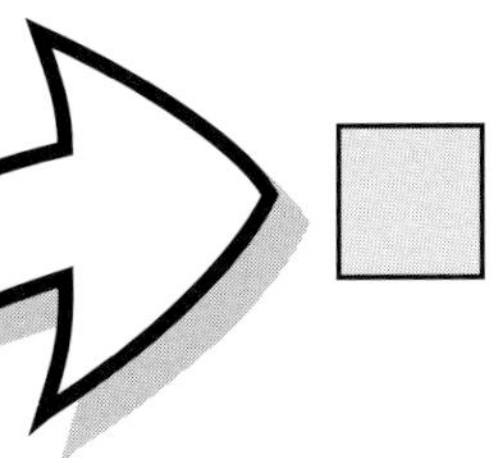

1 cm^2

Since each piece of foil was one square centimeter, Ulrichs was able to find the total surface area of his body.

He then put all of the foil squares on graph paper in the shape of a large rectangle. Since there were 18,360 of these squares, he concluded that his surface area was 18,360 square centimeters, or 18,360 cm^2.

5. Find some possible measurements for the length and width of Timm Ulrichs's rectangle. You don't have to be exact.

6. Find a rectangle with an area of about 18,360 cm^2, whose length could be Timm Ulrichs's height. Is this rectangle $\frac{3}{5}$ of Mr. Ulrichs's square? Explain your answer.

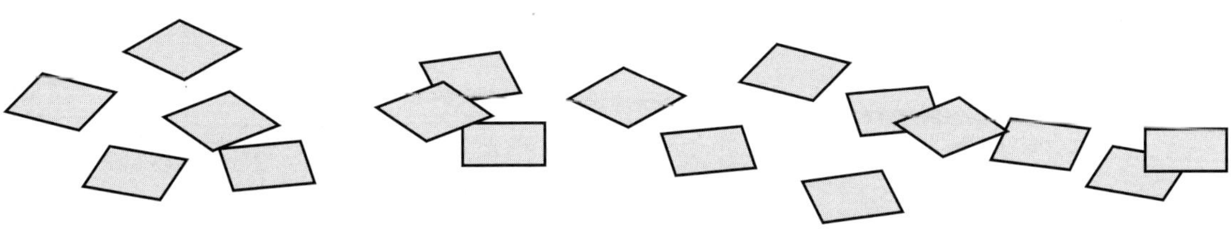

Hands and Body

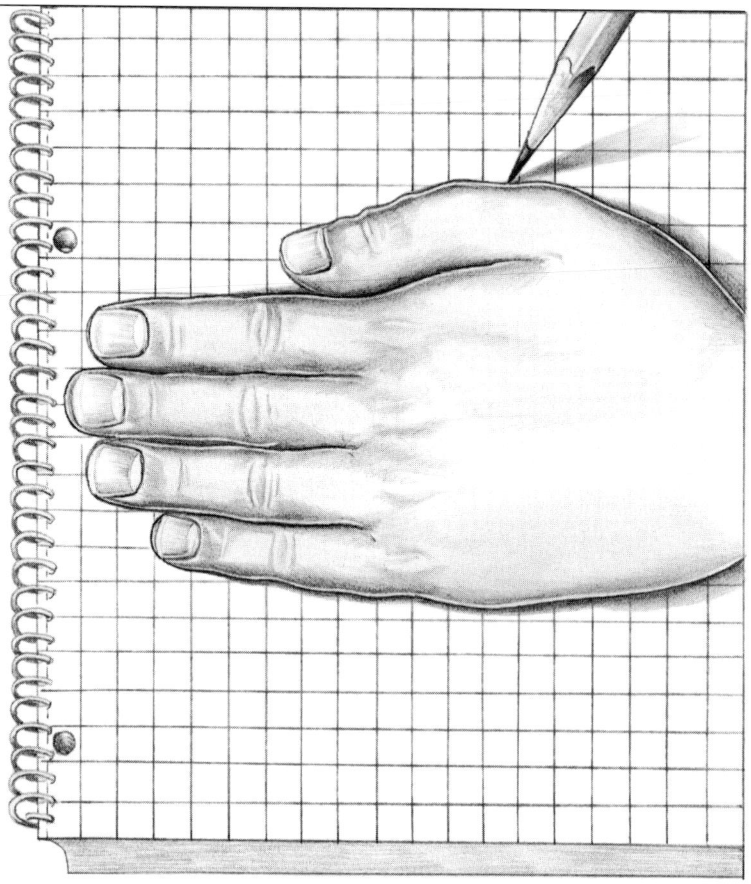

7. Trace your hand on grid paper as shown on the right. Estimate the area of your handprint in square centimeters.

Another method for estimating body surface area is to use the area of a handprint.

About 100 handprints cover the body.

8. Does the handprint method give the same estimate of your body's surface area as Timm Ulrichs's rectangle method? Record the result of the handprint method in your table.

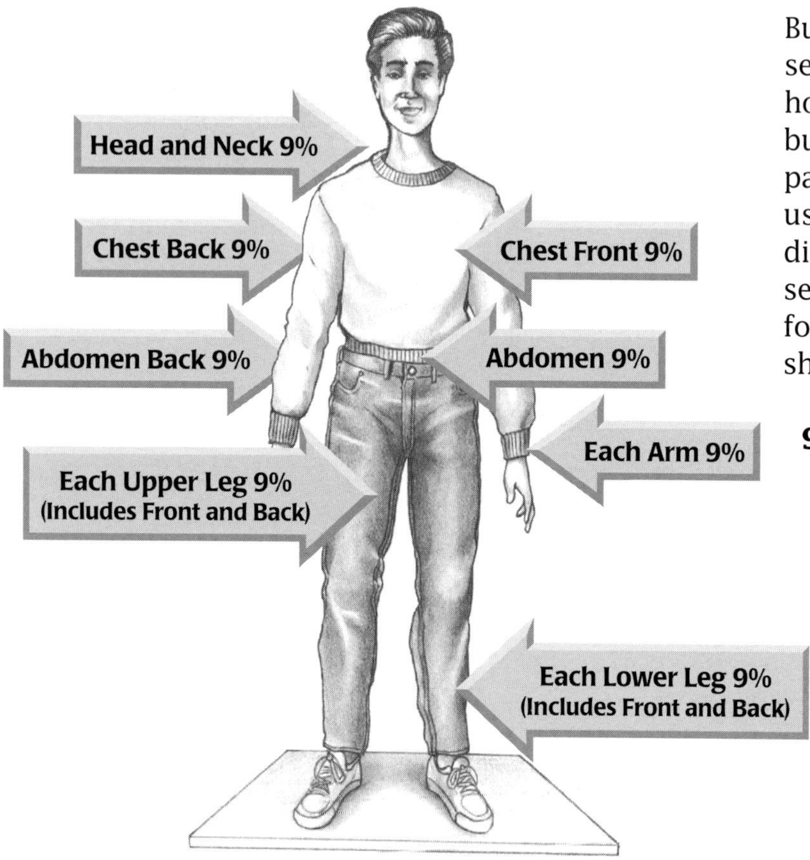

Burns can be very serious. The seriousness of a burn depends on how much of the body has been burned. To estimate the extent of a patient's burns, health care workers use the "rule of nines." This rule divides the body into eleven sections, each of which accounts for 9% of the total surface area, as shown in the picture on the left.

9. a. Think of a way to measure as precisely as possible one of the "rule of nines" sections of your own body surface. Measure the area of this section in square centimeters.

   b. Use the result to calculate your body's surface area. Record your body's surface area in your table from page 13 with your other estimates.

Surface Area by Formula

There are both simple and complex formulas for finding the surface area of a person's body. One simple way is to multiply your height by your thigh circumference (the length around your thigh) and double the answer. The formula for this method is shown below.

height (in cm) × thigh circumference (in cm) × 2 = body surface area (in cm^2)

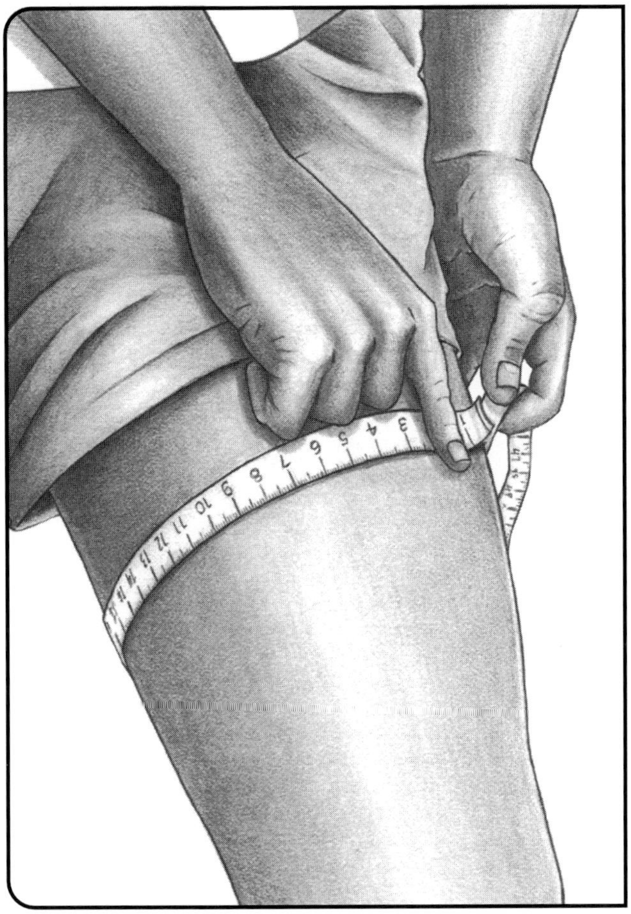

10. Measure your height and thigh circumference in centimeters, and then find your body's surface area using the above formula. Write the result in your Surface Area Estimation Table.

The above formula is based on the formula for the surface area of two cylinders.

11. Explain how the body can be seen as two cylinders of equal height.

12. Would the above formula work for babies? Explain why or why not.

Height, **Weight**, *and* Area

The surface area of a person's body is related to his or her height and weight. Unfortunately, there is no easy formula to find out exactly how these three measures are related.

Pictured on the right is a chart that health care workers often use to find the surface area of a patient's body. The chart is called a *nomogram.*

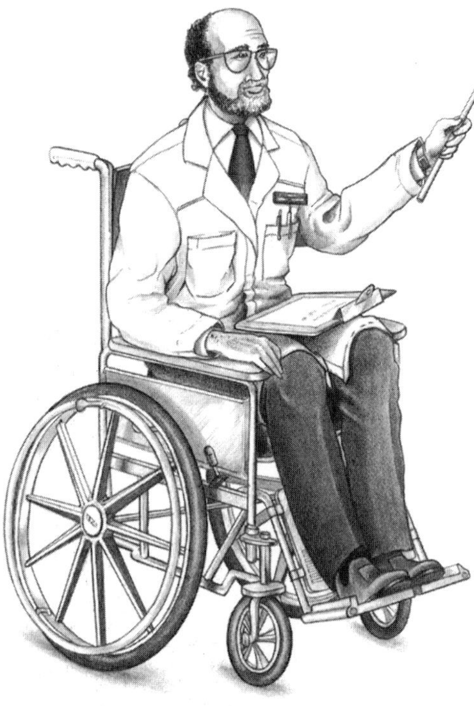

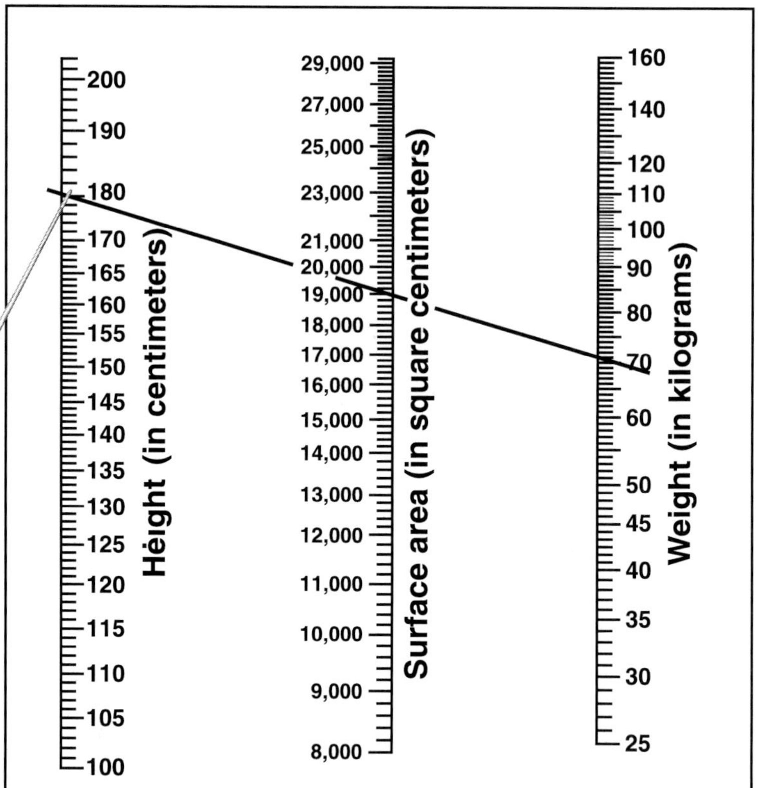

Nomogram for Estimating Body Surface Area from Height and Weight

To use the above nomogram, first find the patient's height (in centimeters) in the left-hand column. Next, find his or her weight (in kilograms) in the right column. Then connect these two points with a line. The surface area of the patient's body is shown where the line crosses the middle scale.

13. **a.** A line for one patient has already been drawn on the nomogram. What information about the patient is indicated by the nomogram?

b. Use the nomogram and a straightedge to estimate your own body's surface area, and then enter the result in your table. (*Note:* 1 kilogram = 2.2 pounds)

EARLY AREAS

In Section A, you saw that body measures were often used to find lengths in the past. At one time, area measures were also based on the human body. Square feet, square furlongs, and square thumbs were all used to measure area.

Area measures used in the United States were based on English measures. For areas, the English used such units as the *rood* and the *acre*. The acre is still used today.

In Section A, you discovered that a furlong is 660 feet long. Also, one square furlong equals 10 acres.

14. How long and how wide might 1 acre be? Express your answer in feet. Make a drawing to explain your answer.

A football field is about 100 yards long by 50 yards wide.

15. Use your answer to problem **14** to calculate how many acres are in a football field.

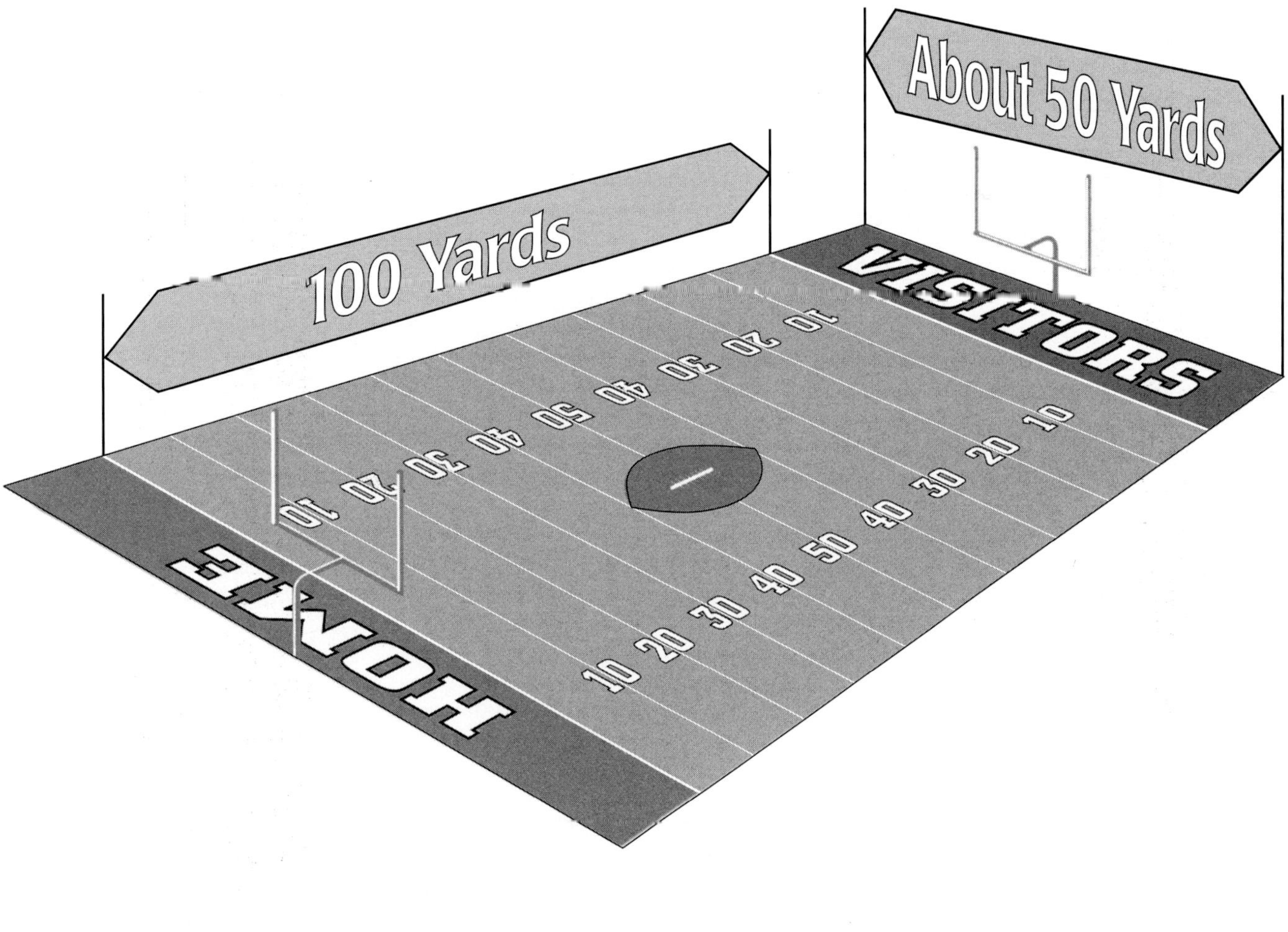

Summary

Surface areas are important for figuring out the amount of material needed to cover something. Body surface areas are used to determine the amount of fluid needed by burn victims.

In this section, you learned about several different methods to find your body's surface area. These methods used squares ($\frac{3}{5}$ of your square), handprints (about 100 handprints), a formula based on the area of two cylinders (height × thigh circumference × 2), and a nomogram.

Summary Questions

16. Write a paragraph or two comparing the different methods you used for estimating body surface area. Which of these methods do you think is most accurate? Why?

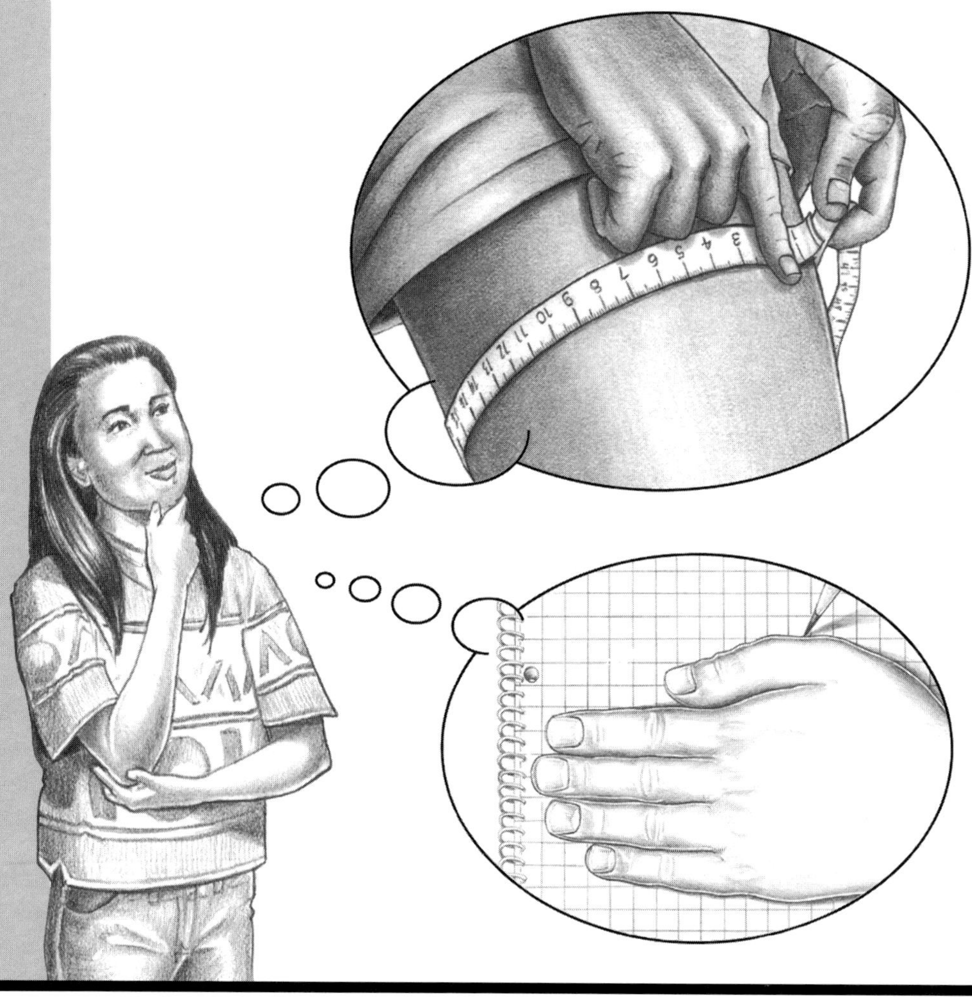

C. VOLUMES

INTRODUCTION

In the last two sections, you measured the lengths and surface areas of different parts of your body. You can also measure, estimate, and calculate volumes. Knowing the volume of a three-dimensional object is useful when you want to describe how much space it takes up.

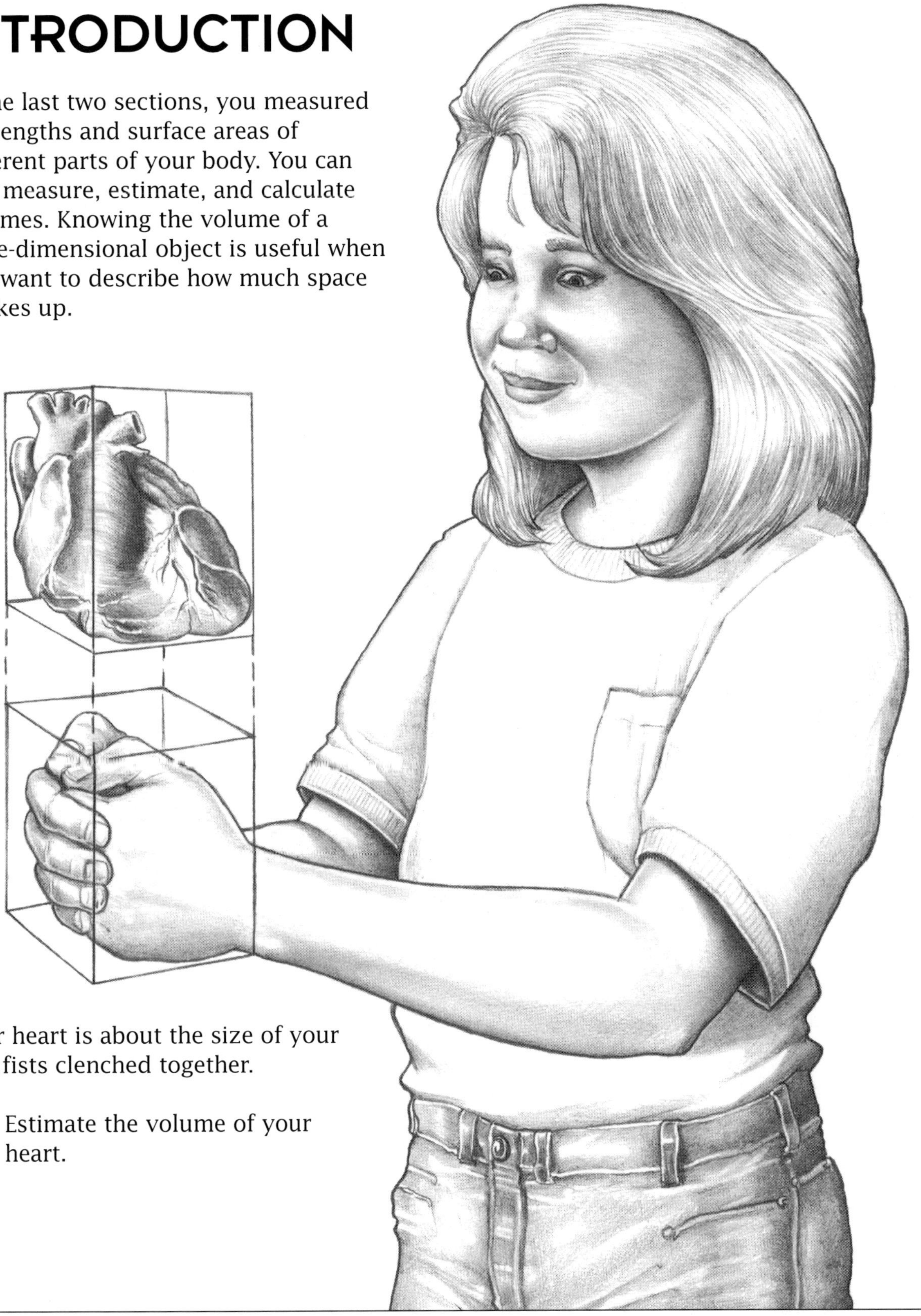

Your heart is about the size of your two fists clenched together.

1. Estimate the volume of your heart.

SOLIDS

2. **a.** You can make a centimeter cube (1 cm^3) by drawing a figure like the one on the right and then folding it up into a cube. The figure is called the *net* of a cube. Use centimeter grid paper to draw a net of a centimeter cube. Cut it out and then fold and tape it to make one cubic centimeter.

b. Are all of the centimeter cubes made by your class enough to fill up one cubic decimeter? (*Note:* there are 10 centimeters in one decimeter.)

c. How many centimeter cubes are needed to completely fill one cubic decimeter?

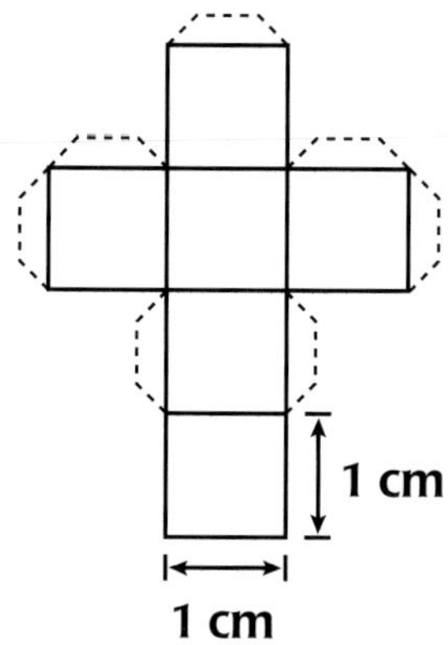

3. Cut the top off a tissue box. How many cubic centimeters are needed to fill it? Explain how you found your answer.

If it takes 32 centimeter cubes to completely fill a box, we say that the volume of the box is 32 cubic centimeters. The volume of larger objects can be expressed in cubic decimeters or cubic meters (in the metric system) or in cubic feet or cubic yards (in the English system).

4. **a.** For what kinds of objects would volume be measured in cubic centimeters? cubic decimeters? cubic meters?

b. Write some statements about how cubic centimeters, cubic decimeters, and cubic meters are related.

Maria is trying to find the volume of the box shown below. The box has been partially filled with cubes.

Maria says, "I can easily find the volume of the box. Since the box measures 8 centimeters by 7 centimeters, I can fit 56 cubes on the bottom layer."

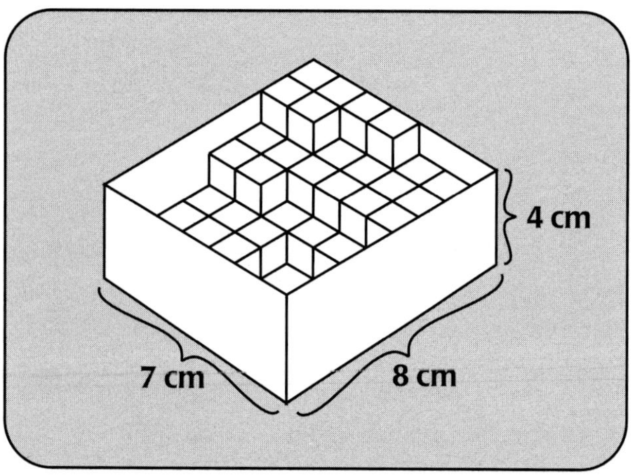

5. Explain what else Maria has to do to find the volume of the box.

6. Jonathan has a fragile vase that he is mailing in a box. To cushion the vase, he bought a box of packing material that says "Will fill a box as big as 6,000 cubic centimeters!" His box is 35 centimeters by 16 centimeters and has a height of 10 centimeters. Will there be enough packing material to fill the box? Show your work.

7. Margaret has four different boxes. Each one has a volume of 18 cubic decimeters. What might the measurements of these boxes be?

8. Find the volume of each item below and describe your strategy.

a.

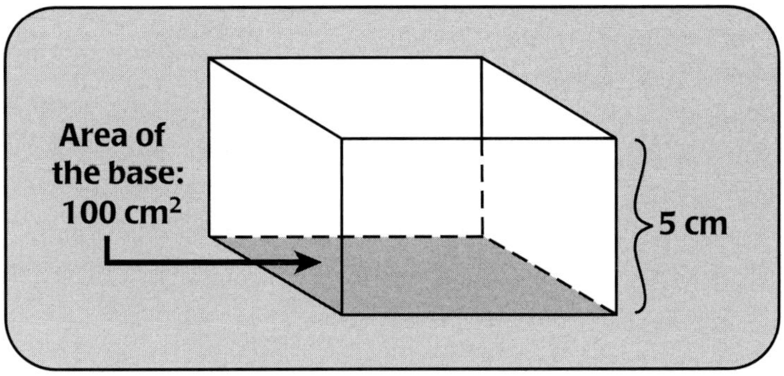

b. a deck of cards

c. a soda can (*Hint:* What is the area of the bottom of the can?)

9. Suppose that you want to determine the volume of your classroom. Can you use the area of one wall to find the volume of your classroom? Find the volume of your classroom.

10. Is it possible to use centimeter cubes (like the one you made on page 22) to find the volume of a soda can? Explain your reasoning.

You can often find the volume of an object by using the following formula:

volume = area of the base × height

11. The above formula does not always give the correct answer for all three-dimensional objects. For example, the volume of the pyramid shown on the right cannot be found by using this formula.

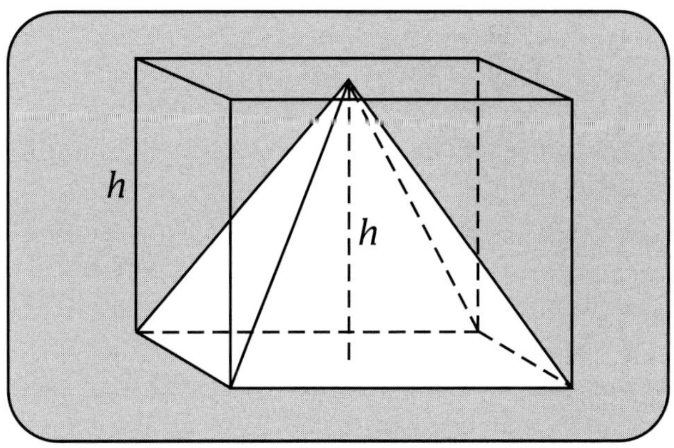

a. Give an example for which the above formula does work and another example for which the formula does not work.

b. Why does the formula give the correct answer for some objects but not for others?

You can use the area of your handprint (Section B, problem **7**) to estimate the volume of your hand.

12. a. Besides the area of your handprint, what other measure(s) do you need to estimate the volume of your hand?

b. Estimate your hand's volume.

LIQUIDS

The units used to measure liquids are different from the units used to measure solids. Pints, quarts, gallons, liters, and milliliters are all typical measures of liquid volume.

Shown on the right are some bottles whose contents are measured with units of liquid volume.

You might find it helpful to know that

- a regular soda can contains about $\frac{1}{3}$ liter.
- a cubic decimeter contains exactly 1 liter of liquid.

TRY THESE!

13. **a.** Estimate the volume (in liters) of the objects shown on the right.

b. Estimate the volume (in liters) of the following objects:

- a small container of juice
- a large cup of soda
- a small cup of coffee
- a gallon of milk

Activity

Measuring the Volume of Your Hand

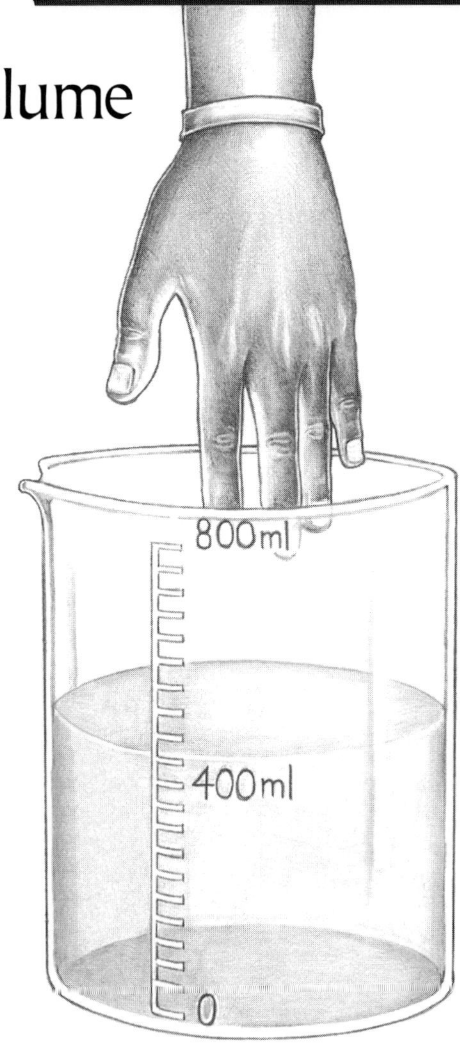

For this activity, you will need a can or beaker that measures in milliliters. It should be big enough so that you can put your hand in it.

Pour water into the can or beaker so that it is about half full. Measure and record the level of the water. Then, put a rubber band on your wrist, and put your entire hand in the water up to the bottom edge of the rubber band. You may want to make a fist. Be sure that your whole hand is underwater.

Measure and record the new water level. The difference between the old level and the new level is the volume of your hand in milliliters.

You now have two estimates of the volume of your hand: one in cubic centimeters (from problem **12**) and the other in milliliters. How are these two measurements related?

14. Make a table listing the two estimates of hand volumes for 10 students. Find a relationship between the two measurements. Drawing a graph might be helpful.

The Volume of Your Body

You could find the volume of your body in the same way that you found the volume of your hand, using liquid measures, but you would probably need a bathtub to do it!

Another way is to use solid units of measure and to think of your body as being made up of cubes

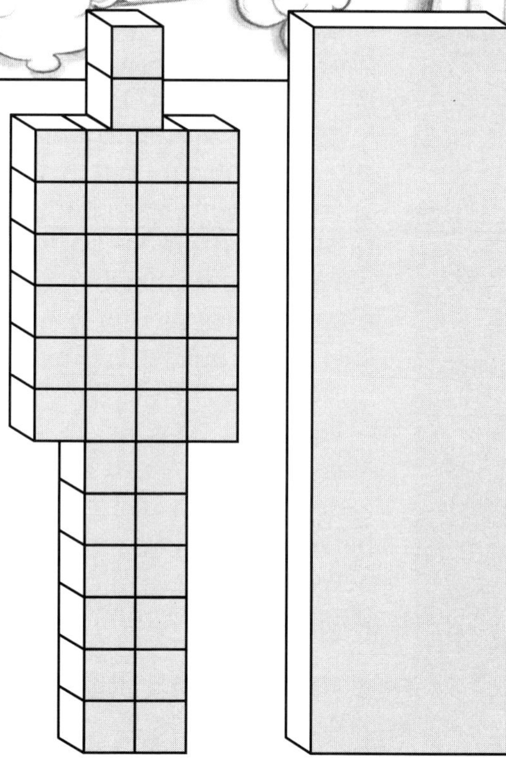

15. Estimate your body's volume by modeling your body with cubes as shown on the right. You will need to decide on a good unit of measure for the cubes.

To estimate the volume of your body, you can also use a block to model your body, as shown on the far right.

The table below contains some information about a girl at ages 10, 13, and 16.

Data for One Girl at Three Ages

Ages (Years)	10	13	16
Height (in dm)	14	16	17
Mass (in kg)	31	52	65
Surface Area (in dm^2)	110	150	175
Volume (in dm^3)	41	67	79

16. Suppose that you want to draw a block that can be used to model the volume of the girl at age 13. Find the width and depth measurements of the block.

Other Measures for Volume

In the past, units of measure for volume were sometimes related to body sizes. It was more common, however, to use measurement tools like the cup as standards for measuring volume.

The ancient Egyptians had the first unit of measure for length: the cubit. The *cubit* is the distance from a person's elbow to the tip of the extended middle finger.

The Egyptians used a standard length called the *royal cubit,* which is 52.4 centimeters, or 20.62 inches.

In the ancient Egyptian system, the volume unit of measure is the *cubic cubit.*

17. a. Use your two arms to get a rough idea of how big a cubic cubit is.

b. About how many "royal" cubic cubits are there in 1 cubic meter?

The cubic fathom is a unit of measure used in the 1800s in Europe to measure volumes of firewood. (Remember that a fathom is the distance between your two outstretched arms.)

18. a. Estimate the amount of firewood that can fit into your cubic fathom.

b. What else can you measure with a cubic fathom? with a cubic cubit?

In the United States, firewood is measured in *cords.* A *face cord* is 4 feet high, 8 feet long, and 16 inches deep. A face cord usually fits in the back of a pickup truck. A *full cord* is 4 feet high, 8 feet long, and 4 feet deep.

19. How much more firewood is in a full cord than in a face cord?

Summary

Volume is important for measuring the sizes of such objects as a box, a drink, or even your hand.

In this section, you measured volumes in two ways: with liquid units of measure (such as liters) and with solid units of measure (such as cubic centimeters).

You also measured the following:

- the volume of an object when you wanted to know how much space it takes up or how much it holds.
- the surface area or area of an object when you wanted to know what it takes to cover it.
- the length of an object when you wanted to know how long or how tall it is.

Summary Questions

20. a. When would it be useful to know the volume of an object?

b. Pick an object and estimate its volume. You may use any units and any measuring method you wish.

21. Linda knows that 1 liter equals "one cubic something." She doesn't remember whether it is 1 cubic centimeter, 1 cubic decimeter, or 1 cubic meter. Explain to Linda which unit of measure equals 1 liter.

D. ANGLES

INTRODUCTION

When designing furniture, cars, computers, and other items, it is sometimes necessary to know the angles that a person can make with his or her arms, ankles, wrists, and head.

1. For what objects might it be important to keep angles in mind?

Ergonomics is important for designing work environments. Designers use ergonomics to determine the placement of office equipment and the size and dimensions of furniture that will create safe, efficient working environments for people. Designers also consider ergonomics when creating new buses and trains for commuters, for example, by placing controls close to the driver.

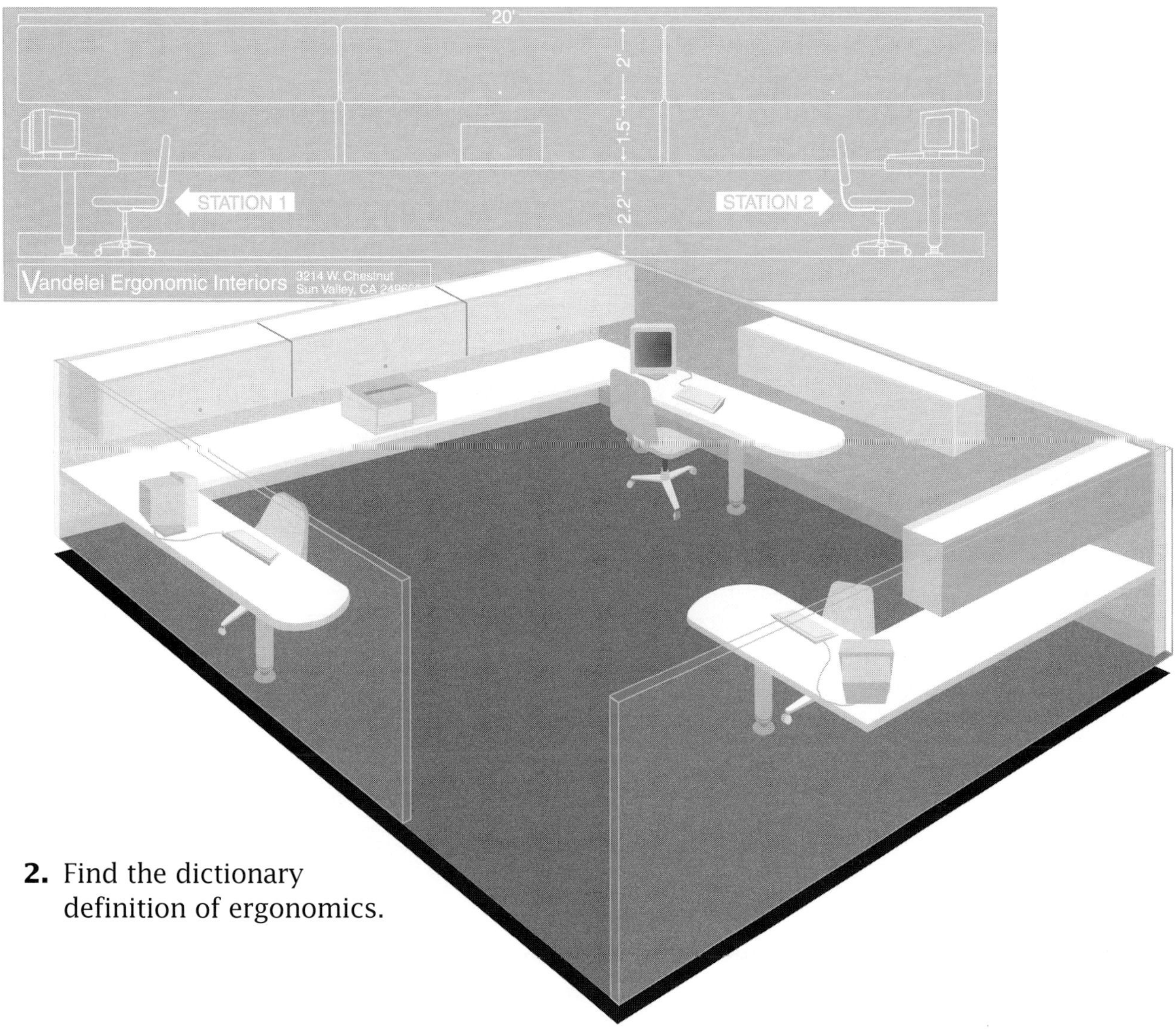

2. Find the dictionary definition of ergonomics.

Activity

Arcs of Movement

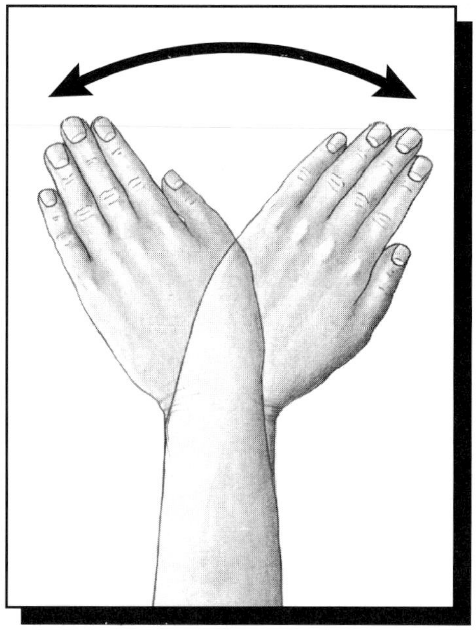

On the right is the "arc of movement" of someone's left hand. The picture shows how far to the left and right the wrist can bend.

You can measure the arc of movement of your own hands.

i. Draw a small × on the bottom of a sheet of paper.

ii. Lay your forearm flat, with your left hand and wrist on the sheet of paper. Put the middle of your wrist over the small ×.

iii. Put a mark on the paper at the top of your middle finger.

iv. With your arm and hand flat on the table, bend your wrist to the right as far as you can without moving your arm. Put another mark on the paper at the top of your middle finger.

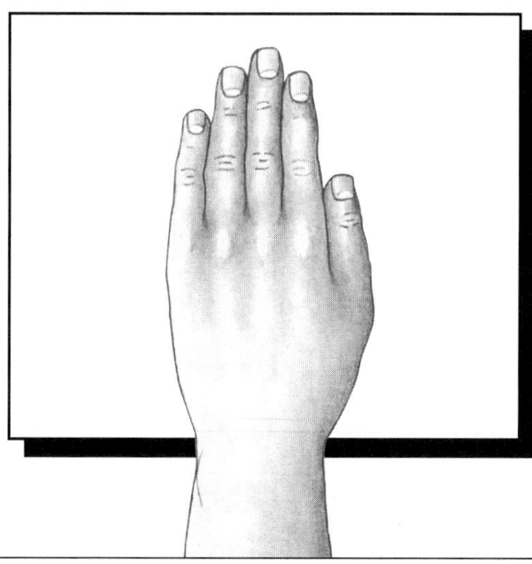

v. Bend your wrist to the left as far as you can without moving your arm. Mark the location of the top of your middle finger.

vi. Draw lines from each of the three finger marks to the small × (the middle of your wrist).

3. a. Use your protractor to measure the angles that your left hand can bend to the right and to the left.

b. What do you think the angles will be for your right hand? Measure the arc of movement of your right hand following the same instructions on page 30 and above.

4. a. Make a table of the right hand angle measurements for all of the students in your class. Describe the results.

b. Why might someone be interested in studying the motion of the wrist joint?

When typing on a computer keyboard, your hands should rest on the keys. In order to reach all of the keys, your hands must bend sideways. If you do a lot of typing, the position in which you hold your hands on the keyboard can cause physical discomfort in your hands, wrists, or forearms.

Studies on ergonomic keyboard design have found that most people prefer a split keyboard. A *split keyboard* is divided into two parts: the left part ends with the keys T-G-B, and the right part starts with keys Y-H-N.

On a split keyboard, the two parts make an angle of about 25°, that is, the lines drawn through T-G-B and Y-H-N make an angle of 25°. The distance between the two parts—or between the keys G and H—should be about 95 millimeters.

5. a. Make a drawing of a split keyboard. You do not have to draw all of the keys, only an outline of the two parts.

b. What is one advantage to using a split keyboard?

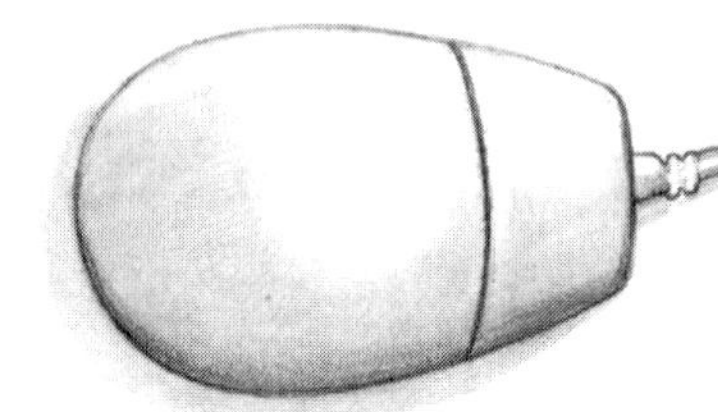

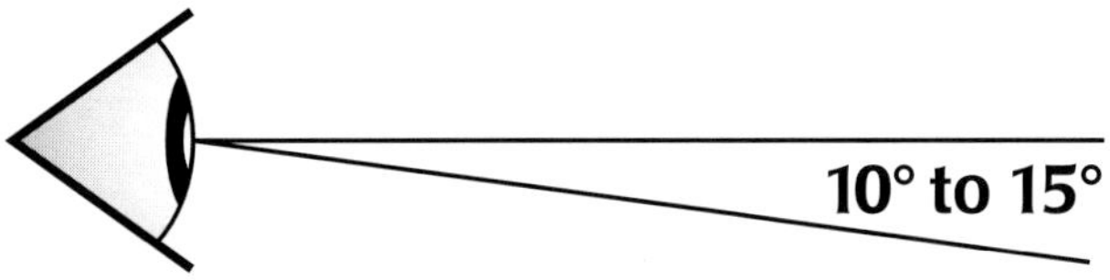

Another important angle in design involves your *line of sight*. If you look straight ahead, your line of sight follows a horizontal line. Your normal line of sight is typically 10° to 15° below the horizontal.

In the best ergonomic computer design, the screen should not be lower than the sight line of 15° below the horizontal.

6. Draw a picture of the situation described above.

You can also use ergonomics to design an easy chair. Some typical recommendations for the design of an easy chair follow.

- The seat should be tilted backward so that you will not slide off. A tilt of 14 to 24 degrees to the horizontal is recommended.
- The angle between the backrest and the seat should be from 105 to 110 degrees, and the angle between the backrest and the horizontal should be from 110 to 130 degrees.

Shown below are two different easy chairs.

7. Check to see if these chairs fit the above recommendations.

8. Based on the above recommendations, design your own easy chair. Draw a side view of your chair on an appropriate scale.

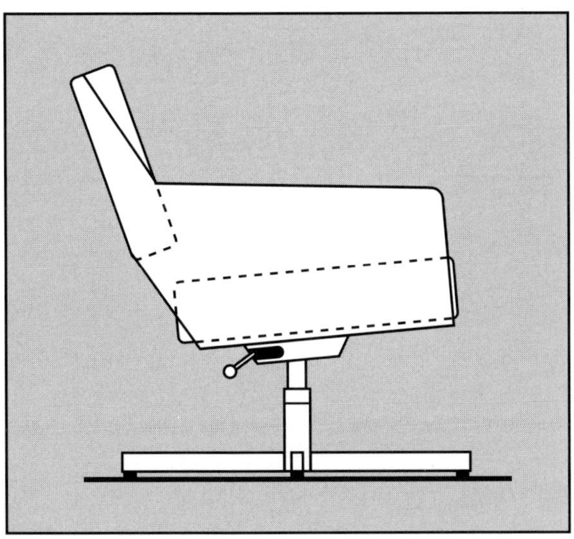

Shown on the left is the side view of an office chair.

9. a. Measure the angles the seat and the backrest make to the horizontal plane. Also measure the angle of the backrest to the seat.

b. Compare your findings to the above recommendations for easy chairs. What conclusion can you make?

Summary

In this section, you investigated the movement for both the right and left hand. You also explored how the placement of a computer screen and the angles between the seat and the backrest of an easy chair can affect your comfort.

Summary Questions

The following text is taken from a book on ergonomics, *Fitting the Task to the Man*, by Etienne Grandjean.

"The arm can rotate through an angle of 250° about its axis . . . of which a half-circle (180°) lies in front of the body, and a further 70° or thereabouts, backwards."

10. a. Read the text carefully and explain what it means.

b. Check to see if your arm can make an angle of 70° backward. First think of a way to measure this angle.

TRY THIS!

Section A. Lengths

1. List some body measures that would be useful to know if you were designing the following:

a. telephones

b. children's beds

c. kitchen cabinets

2. List all of the units of (length) measure that you know that are related to the human body. Explain the meaning of each measure.

3. Which of the measures in problem **2** would you use to measure the following:

a. the height of a door

b. the length of a city block

c. the length of a paper clip

4. Find the U.K. shoe size and the foot length in stitches for the following foot lengths. (Remember the rule: In the U.K. System, shoe sizes are based on the number of stitches. The first 25 stitches are not counted in adult shoe sizes.)

a. 10 inches

b. $9\frac{1}{3}$ inches

c. $8\frac{2}{3}$ inches

5. Which would you use to express the following distances—length or time? Explain.

a. the distance from your home to school

b. the distance from your city to New York City

c. the distance from start to finish of a hiking trail

Section B. Areas

1. Estimate the surface area of the following objects:

a. a basketball

b. a book

c. a cereal box

2. For which object in problem **1** was it easiest to find the surface area? Why?

In class, you found your surface area by thinking of your body as two cylinders of equal height.

3. a. Find another formula for finding the surface area of your body, but, this time, think of your body as one cylinder.

b. Find your body's surface area using the new formula.

4. Enrique lives on a 15-acre farm.

a. About how many football fields is this?

b. What can be the length and width of this farm?

Section C. Volumes

1. List all of the units of measure for volume that you know and explain how they relate to each other.

2. Approximately how many cubic centimeters does it take to fill the following:

a. a cereal box

b. a pack of gum

3. Bill purchased 3,000 cubic centimeters of potting soil to plant flowers in a flower box. What is the largest flower box he can fill with this amount of soil?

4. Find the volume in cubic centimeters of the following.

a.

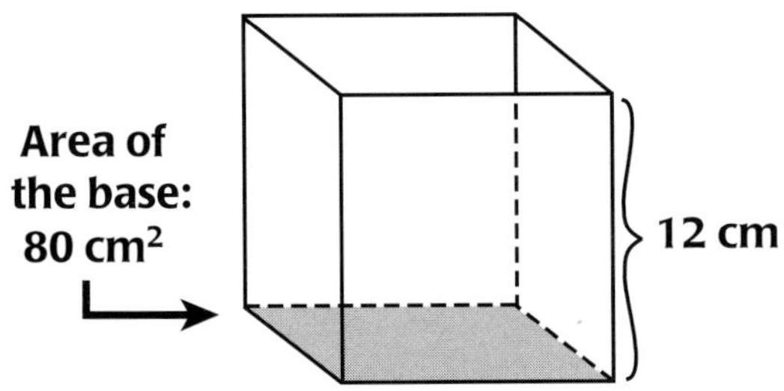

b.

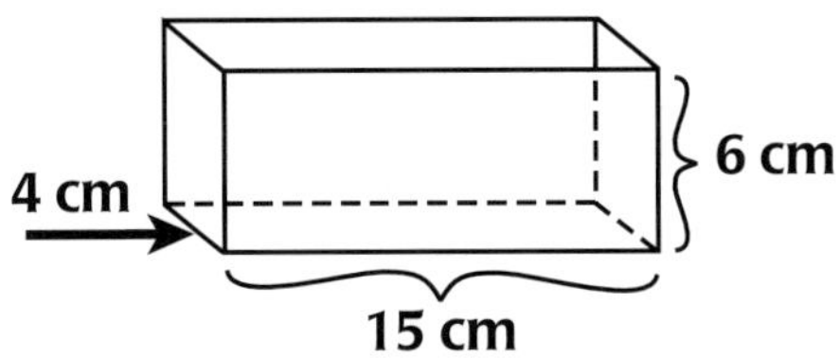

5. Estimate the volume in liters of the following:

a. a bathtub

b. a quart of milk

c. a bottle of shampoo

6. a. Estimate your volume by modeling your body with a block.

b. What are the dimensions that you used for your block?

c. Which method do you think is more accurate for finding your volume—modeling your body with cubes like you did in class or modeling your body with a block? Explain.

D. Angles

1. Draw a side view of a chair so that the angle between the seat and the backrest is 115° and the angle between the seat and the horizontal is 7°.

2. Below is a side view of a computer screen and a keyboard on a desk.

a. Measure the angle of the keyboard to the desk.

b. Measure the angle of the monitor to the horizontal plane of the desk.

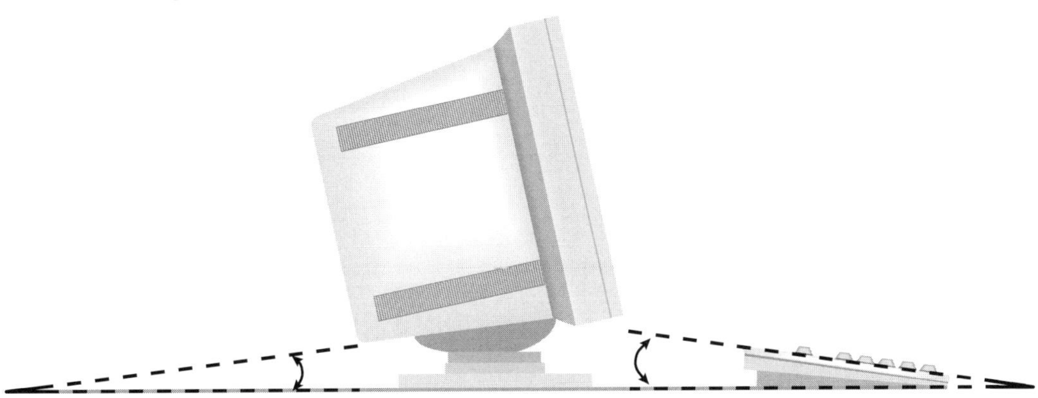

CREDITS

Cover

Design by Ralph Paquet/Encyclopædia Britannica Educational Corporation.

Collage by Koorosh Jamalpur/KJ Graphics.

Title Page

Phil Geib/Encyclopædia Britannica Educational Corporation.

Illustrations

1, 2 (a, b, c, e, f) Phil Geib/Encyclopædia Britannica Educational Corporation; **2 (d), 3** Tom Zielinski/Encyclopædia Britannica Educational Corporation; **4, 6 (top)** Phil Geib/Encyclopædia Britannica Educational Corporation; **6 (bottom)** Tom Zielinski/Encyclopædia Britannica Educational Corporation; **7–8, 9 (top)** Phil Geib/Encyclopædia Britannica Educational Corporation; **9 (bottom)** Phil Geib and Tom Zielinski/Encyclopædia Britannica Educational Corporation; **10–13** Phil Geib/Encyclopædia Britannica Educational Corporation; **14** Phil Geib and Tom Zielinski/Encyclopædia Britannica Educational Corporation; **15, 16 (top)** Phil Geib/Encyclopædia Britannica Educational Corporation; **16 (bottom), 17–18** Phil Geib and Tom Zielinski/Encyclopædia Britannica Educational Corporation; **19–21, 24 (top)** Phil Geib/Encyclopædia Britannica Educational Corporation; **24 (bottom)** Phil Geib and Tom Zielinski/Encyclopædia Britannica Educational Corporation; **25–29** Phil Geib/Encyclopædia Britannica Educational Corporation; **30–31** Phil Geib and Tom Zielinski/Encyclopædia Britannica Educational Corporation; **32–33** Phil Geib/Encyclopædia Britannica Educational Corporation.